Connect with Nature books

THE SUN AND ME

a story about the daily body clock

by Suzanne Mathis McQueen

Illustrations by Pumudi Gardiyawasam

Biology ~ Nature ~ Health
Paperback ISBN: 978-0-9858955-2-5

Special thanks to:

Ask Janis Editorial Services
Red Letter Editing
Pumudi Gardiyawasam, illustrator extraordinaire!
Ms. Galaxy, Jessica Vineyard
Rocco and Jazzi (the dogs)
Patrick Wagesreiter, permission for Patrick Hand font
Blaine and Erin Moore of Apex Authors

Superstar Beta Book Readers:
Myan McQueen, Louise Abel, Rosemary Mathis LaBonte, Linda Parker, Sofia Henry, Sara Parker, Mary McJilton Benarth, Linda Jolly, Cynthia Salbato, Laura Knapp, Jane Almquist, Gretchen Lee, Jeanie Vollrath, Feather Gilmore, Judith Clark, June Kenney-Clark, Eliza Kenney-Clark, Tracie Smith, Mark Siino, and Aurora Perez.

P.K. Halinan,
for his kindness, encouragement, and sage advice about writing children's books when we'd run into each other on morning walks.

And appreciation to Amy Bammel Wilding, author, leader, and sister/friend for writing a line in a blog that sparked the idea for this book series.

The Sun

I told the Sun that I was glad,
I'm sure I don't know why
Somehow the pleasant way he had
Of shining in the sky,
Just put a notion in my head
That wouldn't it be fun
If, walking on the hill
I said "I'm happy" to the Sun.

by John Drinkwater

For my sister, Ro, and my brother, Kim

Here's to our fun times while playing outside under the perfect Southern California Sun.

The Sun, the Moon, and the Seasons
are magical timekeepers!

Our bodies follow the Sun's daily clock,
the Moon's monthly clock, and the Four Seasons' yearly clock.

This is the story of the Sun's clock and me!

The Sun wakes us up in the morning.

Each day, it enters through the
east gate of the sky and says, "Hello!"

It brings us the gifts of light, color,
and warmth.

The Sun keeps us healthy and
helps the plants to grow.

Throughout the day, the Sun travels across the sky.

As I go about my day,
my energy follows the
light of the Sun.

In the morning, the Sun helps me get ready for the day.
My energy builds as I make my bed, eat breakfast,
and brush my teeth.

I'm all set now and feel excited!

I hear the Sun call out to me:

"Hey, you! Come out
to explore and to learn,
and to work and to play!"

So I do!

My friends in the neighborhood are already outside!
I run to join them at the big oak tree.

We pretend to be pirates
looking for hidden treasure.

We find acorns, tiny leaves, funny-shaped rocks, and
sometimes coins or sparkly things.

Here's a secret:
My dad helped us paint and glue things to
a cardboard box to make a treasure chest!

We also look for slimy worms to hold
and ladybugs to chat with.

In the middle of the day
the Sun is at its highest in the sky.

I stretch my arms out and
close my eyes tight so that they don't get
burned by the Sun's strong rays.

My energy feels as **BIG** as the Sun!

I don't want to stop playing!

So I don't (except to eat lunch).

Life is really fun.

The Sun keeps traveling into the late afternoon.
It's now on its way back home.
It's been a busy day!

Like me, the Sun must be tired.

I hear my parents calling
me to come inside.

I say goodbye to
my friends.

As the Sun drops down through the west gate of the sky,
it leaves us with a gift.
It paints the clouds with pink and orange colors!

The Sun calls out to me one more time.

"Goodbye! Thank you for a fun day!
Be sure to get a good night's sleep so that
we can play again tomorrow.

"Oh! And say hello to the Moon for me!"

So I say:

"I will! Thank you, Sun, for waking me up this morning and helping me to play BIG during the day!

"I'm tired now, and like you, I'm going home.

"I promise to get a good night's sleep so that we can play again tomorrow."

It's dark now.

The Moon appears through the east gate of the sky and says "Hello!"

Sometimes it passes through the gate when the sky is still light and I'm not looking.

(The Moon is kind of sneaky that way.)

Whether it is day or night
and whether I can see it or not,
the Moon travels across the sky, just like the Sun does.

But I'm too sleepy to stay up and
play with the Moon tonight.

I crawl under my blankets, get cozy, and close my eyes
so that I can have another good day tomorrow.

Good night!

Fun facts for the family...

The Sun is like a big clock in the sky that helps your body keep track of the time each day.

Your body needs a lot of activity and lots of sleep to stay healthy.

When the Sun is in the east gate, it is telling your body to get up in the morning and be active.

When the Sun is in the west gate, it is telling your body to go to bed to sleep deeply.

The Sun even guides you through at least FOUR different kinds of energy in one day!

And then it does the same thing the next day, and the next day, and the next day!

This is called a "body clock", or a "body rhythm."

There are many body rhythms inside of you.

The one that follows the Sun's light each day is called a "24-hour body clock" or a "circadian rhythm."

sur-KAY-dee-in means "around a day."

RITH-um means "movement in time."

Questions

1. What kind of energy do you have in the morning?

2. What kind of energy do you have in the middle of the day?

3. What is your energy like in the evening after eating?

4. What does your energy feel like when you crawl into bed?

5. Do you fall asleep easily?

6. Do you feel more energized on sunny days or cloudy days?

Sleeping

If you have trouble getting to sleep at night or getting up in the morning, try one or more of these ideas:

- Go to bed and get up at the same time every day.
- Turn off any lamps near your bed and the overhead light in your room. Night lights are nice.
- Read from a book before going to sleep instead of looking at a computer, phone, or TV. Those lights make your body think the Sun is still out.
- Look to see if the Moon is full. If so, say "Hello!" and then close the curtains.

Project Ideas

Search online to learn how to do any of these things. It's fun and easy to find ideas!

- Make some Sun-shaped cookies.
- Draw the Sun with different faces.
- Plant some morning glory flowers or watch a video of them. Notice how they open or close with the sunlight!
- Make a sundial!

Have fun with the Sun!

A Note from the Author

The Sun and Me is the first book in a series that introduces children and their people to the connection between Nature's rhythms and their own body rhythms. Learning to synchronize their daily, monthly, and yearly habits with the Sun, the Moon, and the Seasons may help families achieve ideal lifelong health and happiness!

Please note that the science of the Earth and Moon's orbits differs from how humans experience them. For instance, the Sun never sleeps, but we can't see it when it is busy lighting the other side of the world. Yet, the Sun continues to affect us! The darkness it leaves behind triggers our bodies to get sleepy.

This book tells *that* story.

Aligning with the Sun creates a daily pattern and routine that supports balanced *physical* and *mental* well-being.

I encourage your family to notice how the Sun's patterns affect your decision-making and to incorporate the conversation into your daily language (but never look straight at the Sun — your eyes will get damaged!). Words not usually capitalized are done so because they are considered the series' main characters.

The Sun and Me, The Moon and Me, **and** ***The Seasons and Me*** teach body rhythm information to kids of all genders. The *Connect with Nature* series continues with books to help those with girl bodies to know and love the superpowers of their monthly hormonal rhythms beginning at adolescence, based on the *4 Seasons in 4 Weeks* cycle-tracking method.

Suzanne Mathis McQueen

Suzanne is the author of the book *4 Seasons in 4 Weeks*.
She is a Nature nerd who hugs trees and
chats with ladybugs at her home in Southern Oregon.
She likes both sunny and cloudy days, depending on the time of year.

Do you have a favorite
time of the day to
play with the Sun?

Have you read all of these books by Suzanne Mathis McQueen?

The Connect with Nature Series

For children & their people about body clock awareness for health & happiness!

and...Coming Soon!

A coming-of-age picture book for 7-11- year-old girls!

For women, couples, and mature teens:

Based on science, the *4 Seasons in 4 Weeks* cycle-tracking method identifies an approximate 28-day female body clock and breaks it down into four non-technical guided weeks for optimal decision-making in health, business, and love.

Includes women's history, sacred geometry, tools and self-care guidance for each cycle week, information for men, and guidance for not-cycling women as well. Explicit content with 300 pages of full color, gorgeous artwork, and color-coded sections.

Reviews are appreciated...

If you enjoy reading *The Sun and Me* and believe other kids (and their people) would also like this book, please leave a review on your favorite review site! Reviews help readers find good books and independent publishers (like us!) to stay in business.

Thank you!

Stay in touch...

Scan this code or type in the URL below to find coloring pages, contact info, merch, and other bits of fun:

4s4w.com/connect-with-nature-books/

Made in the USA
Coppell, TX
04 November 2024

39616800R00024